EQUATIONS AND INEQUALITIES

Alessio Mangoni

©2021 Alessio Mangoni. All rights reserved.
ISBN: 9798512350348

DR. ALESSIO MANGONI, PHD
Scientist and theoretical particle physicist, researcher on high energy physics and nuclear physics, author of many scientific articles published on international research journals, available at the link:

http://inspirehep.net/author/profile/A.Mangoni.1

https://www.alessiomangoni.it

I edition, May 2021

Contents

Contents   5

Introduction   9

1   Numeric sets   11
    1.1   Natural numbers 11
    1.2   Integer numbers 12
    1.3   Rational numbers 12
    1.4   Irrational numbers 12
    1.5   Real numbers 13

2   The equations   15
    2.1   Definition 15

	2.2	Types	16
	2.3	Principles of equivalence	17

3 The first degree equations — 21

- 3.1 Integer equations — 21
- 3.2 Integer literal equations — 24
  - 3.2.1 Exercise — 25
  - 3.2.2 Exercise — 27
- 3.3 Systems of equations — 28
- 3.4 Cramer's method — 32
  - 3.4.1 Exercise — 35

4 The second degree equations — 39

- 4.1 Generic form — 39
- 4.2 Solution method — 41
  - 4.2.1 Exercise — 43

5 The fractional equations — 45

- 5.0.1 Exercise — 46

6 The intervals — 49

	6.1	Definition	49
	6.2	Types	50

7 Numerical inequalities — 53

	7.1	Definition	53
	7.2	Property	55

8 Inequalities — 59

	8.1	Definition	59
	8.2	Types	60
	8.3	Principles of equivalence	61

9 The first degree inequalities — 65

	9.1	Integer inequalities	65
	9.2	Integer literal inequalities	69
		9.2.1 Exercise	70
		9.2.2 Exercise	72
	9.3	Systems of inequalities	74
		9.3.1 Exercise	77

10 Sign of products or ratios — 83

10.1 Sign of a product	83
10.1.1 Exercise	88
10.2 Fractional inequalities	91
10.3 Fractional literal inequalities	95

11 The second degree inequalities   97

11.1 Generic form	97
11.2 Solution method	98
11.2.1 Exercise	102

12 The irrational equations   105

12.1 First case	105
12.2 Second case	107
12.2.1 Exercise	108

13 The irrational inequalities   109

13.1 First case	109
13.2 Second case	112
13.2.1 Exercise	113
13.2.2 Exercise	118

Introduction

Equations and inequalities are one of the basic building blocks for the study of mathematics and all related disciplines.

The topics covered are: numerical sets, intervals, inequalities, equations and inequalities of first degree, integer, fractional, literal, second degree, irrational, systems of equations or inequalities, Cramer's method, study of the sign of products and ratios.

After each topic we propose some solved exercises to better fix the concepts.

Numeric sets

1.1 Natural numbers

The set of natural numbers is indicated with the symbol $\mathbb{N}$ and includes numbers

$$\mathbb{N} = \{0, 1, 2, 3, 4, 5, 6, \cdots\}.$$

Generally if you want to exclude the number zero you can use one of the following symbols

$$\mathbb{N}_0, \qquad \mathbb{N} \smallsetminus \{0\}.$$

1.2 Integer numbers

The set of integers is indicated with the symbol $\mathbb{Z}$, it includes natural numbers and all its opposites, i.e.

$$\mathbb{Z} = \{0, \pm 1, \pm 2, \pm 3, \pm 4, \pm 5, \pm 6, \cdots\}.$$

1.3 Rational numbers

The set of rational numbers is indicated with the symbol $\mathbb{Q}$ and includes all numbers that can be written as the ratio of any two integers (which without loss of generality can be considered mutually prime)

$$\mathbb{Q} = \left\{\frac{p}{q},\ p, q \in \mathbb{Z}\right\}.$$

1.4 Irrational numbers

The set of irrational numbers is indicated with the symbol $\mathbb{I}$ and includes all numbers that cannot be written as the ratio between two integers mutually prime. This set also includes the roots of integers that do not belong to $\mathbb{Q}$, such as $\sqrt{2}$ or $3^{0.11}$.

1.5 Real numbers

The set of real numbers is denoted by the symbol $\mathbb{R}$ and includes all rational numbers and all irrational numbers, i.e.

$$\mathbb{R} = \mathbb{Q} \cup \mathbb{I}.$$

The equations

2.1 Definition

An equation consists of the equality between two generic expressions, containing an unknown to be found, usually denoted by the letter x.

An example of an equation is the following

$$2x + 3 = 4 - x.$$

In an equation, the "=" symbol appears, dividing the two members of the equation. The expression $2x + 3$ is called the first member, while the expression $4 - x$ the second member.

2.2 Types

The equations can be grouped as follows:

- **1st degree equation**

  An equation is said to be of the first degree or linear if the unknown appears only at the first degree (x);

- **2nd degree equation**

  An equation is said to be of the second degree if the maximum power to which the unknown is raised is two (x^2);

- **equation of degree $n-$ th**

  An equation is said of degree n-th if the maximum power to which the unknown is raised is n (x^n);

- **numerical equation**

  An equation is called numerical if no other letters appear beyond the unknown;

- **literal equation**

  An equation is said to be literal if other letters appear besides the unknown;

- **integer equation**

  An equation is said to be integer if the unknown appears only in the numerator of any fractions;

- **fractional equation**

  An equation is said to be fractional if there are fractions and the unknown also appears in the denominator.

Obviously not all these groups are independent, you can have, for example, a second degree equation that is also literal.

2.3 Principles of equivalence

Two equations are called **equivalent** if they have the same solutions.

In fact, when we solve an equation, we use rules to transform it gradually into simpler equations, but still equivalent (otherwise we would not get the solution of the initial equation, from which we started).

The simplification steps must comply with the rules contained in the following equivalence principles:

- **First principle of equivalence**

  Given an equation, adding or subtracting any expression (or number) to both sides gives an equivalent equation. In symbols we write:

  $$f(x) = a \implies$$
  $$f(x) + c(x) = a + c(x), \quad \forall\, c(x);$$

- **Second principle of equivalence**

  Given an equation, multiplying or dividing

both sides by any non-zero expression (or number) yields an equivalent equation. In symbols we write:

$$f(x) = a \implies$$
$$f(x) \cdot c(x) = a \cdot c(x), \ \forall \, c(x) \neq 0.$$

Note that the first principle is what justifies "carrying" a term from one member to another by changing its sign.

The second principle, on the other hand, justifies changing the sign of all the terms of an equation.

The first degree equations

First degree equations are also called linear equations.

3.1 Integer equations

An example of an integer first degree equation is the following

$$5(2x - 1) + 3 = \frac{3x}{4} + 3x.$$

To solve it we use the second principle of equivalence and multiply both sides by the denominator, i.e. 4, so as to eliminate the fraction

$$(5(2x-1)+3) \cdot 4 = \left(\frac{3x}{4} + 3x\right) \cdot 4,$$

from which

$$20(2x-1) + 12 = 3x + 12x$$

We eliminate the parentheses, calculating the product, and add the similar terms

$$40x - 20 + 12 = 3x + 12x,$$

from which

$$40x - 8 = 15x.$$

Thanks to the first principle of equivalence, we add the expression

$$-15x + 8$$

to both members, getting

$$40x - 8 - 15x + 8 = 15x - 15x + 8,$$

from which

$$40x - 8 - 15x + 8 = 15x - 15x + 8$$

and

$$25x = 8.$$

Finally, thanks to the second principle we divide both members by the coefficient of x, i.e. 25, obtaining the solution of the starting equation:

$$x = \frac{8}{25}.$$

In general, to solve a linear integer equation one must perform the simplification steps necessary to obtain other equivalent equations, using the two principles of equivalence, trying to isolate the unknown in the first member, until the solution is obtained.

3.2 Integer literal equations

The equations that contain other letters besides the unknown are said to be literal. These letters are sometimes called parameters and must be considered to all intents and purposes as if they were fixed numbers, without however knowing their value.

It is necessary to have a discussion on the parameters to solve the equation and to be able to correctly use the principles of equivalence or general rules. For example, consider the literal integer linear equation

$$a \cdot x = b,$$

where x is the unknown and a and b are real parameters. In this case, it might be tempting to isolate the unknown x to the first member by dividing both members by a, however it is worth considering the case where a is zero. It is necessary to perform the so-called discussion of the literal equation, which leads to the following scheme:

$$\underbrace{a = 0}_{\underbrace{0 \cdot x = b}_{\underbrace{b = 0}\;\;\underbrace{b \neq 0}}} \qquad \underbrace{a \neq 0}_{x = \frac{b}{a}}$$

$$\forall x \in \mathbb{R} \qquad \nexists x \in \mathbb{R}$$

3.2.1 Exercise

Solve the following literal integer equation with the parameter k

$$(3 - k)(2x + 1) = k(x - 1) + 2 - kx\,.$$

Solution

We start with some calculations

$$3(2x+1) - k(2x+1) = kx - k + 2 - kx,$$
$$6x + 3 - 2kx - k = kx - k + 2 - kx,$$
$$6x + 3 - 2kx - k = -k + 2,$$

we take all the terms containing the unknown x to the first member and the others to the second member, using the first equivalence principle once or more,

$$6x - 2kx = -k + 2 - 3 + k,$$

we add the similar terms and collect the first member x,

$$(6 - 2k)x = -1.$$

Now we need to discuss the parameter, in fact we cannot divide by $(6-2k)$ if this were zero, so we have the following scheme

$$\underbrace{6 - 2k = 0}_{\underbrace{k = 3}} \qquad \underbrace{6 - 2k \neq 0}_{\underbrace{k \neq 3}}$$

$$\underbrace{0 \cdot x = -1}_{\nexists x \in \mathbb{R}} \qquad x = -\frac{1}{6-2k}$$

3.2.2 Exercise

Solve the following literal integer equation with the parameter k

$$5x + k - k(2 - x) + 1 = 2k - x\,.$$

Solution

We calculate the product to remove the brackets

$$5x + k - 2k + kx + 1 = 2k - x\,,$$

and add the similar terms

$$5x - k + kx + 1 = 2k - x\,.$$

We separate the terms that contain the unknown, isolating them at first member,

$$5x + kx + x = 2k + k - 1\,,$$
$$6x + kx = 3k - 1\,,$$

from which

$$(6 + k)x = 3k - 1\,,$$

Finally, let's discuss the parameter

$$\underbrace{6 + k = 0}_{} \qquad \underbrace{6 + k \neq 0}_{}$$
$$\underbrace{k = -6}_{} \qquad \underbrace{k \neq -6}_{}$$
$$\underbrace{0 \cdot x = -19}_{} \qquad x = \frac{3k-1}{6+k}$$
$$\nexists x \in \mathbb{R}$$

3.3 Systems of equations

A system of equations is made up of several equations and solving it means finding common solutions to all equations. Consider, as an example, the system of two equations with two unknowns given by

$$\begin{cases} 2x + 3y = 0 \\ y = x + 1 \end{cases}.$$

One of the techniques for solving a system of equations is by substitution. A variable is obtained as a function of the others and is substituted in all the other equations of the system. In this case we already have the y variable as a function of the x (second equation of the system) so we can replace it in the first equation, leaving the second unchanged

$$\begin{cases} 2x + 3(x + 1) = 0 \\ y = x + 1 \end{cases}.$$

By focusing our attention on the first equation we can write

$$\begin{cases} 2x + 3x + 3 = 0 \\ y = x + 1 \end{cases} , \quad \begin{cases} 5x + 3 = 0 \\ y = x + 1 \end{cases} ,$$

$$\begin{cases} 5x = -3 \\ y = x + 1 \end{cases} , \quad \begin{cases} x = -3/5 \\ y = x + 1 \end{cases} ,$$

getting the value of x. To find the value of y we now use the second equation, replacing the newly found value of x

$$\begin{cases} x = -3/5 \\ y = -3/5 + 1 \end{cases} , \quad \begin{cases} x = -3/5 \\ y = 2/5 \end{cases} .$$

Therefore the solution of the system is the pair of values

$$x = -\frac{3}{5}, \qquad y = \frac{2}{5}.$$

THE FIRST DEGREE EQUATIONS

The main methods for solving a system of n equations in n unknowns are the following:

- **substitution method:**
  this method consists in obtaining a variable (as a function of all the other possible ones) from an equation of the system and replacing it in all the other equations. In practice, we pass from a system of n equations in n unknowns to a system of $(n-1)$ equations in $(n-1)$ unknowns;

- **comparison method:**
  this method consists in equating the second members of two equations having the first equal members (it is not among the preferred methods);

- **Cramer's method:**
  this method is very powerful, but it applies only to linear systems, i.e. in those systems

where all the equations are linear (first degree in the various unknowns).

3.4 Cramer's method

We present here the Cramer's method for solving systems of linear equations.

Consider the following linear system with two equations and two unknowns

$$\begin{cases} a_1 x + b_1 y = c_1 \\ a_2 x + b_2 y = c_2 \end{cases},$$

this system is written in **normal form**, i.e. in each equation we have as the first member all the terms with the unknowns x and y ordered with the same logic and the so-called known terms on the second member, i.e. all expressions that do not contain the unknowns. The coefficients of the unknowns can be put in the form of a matrix in this way

$$A = \begin{pmatrix} a_1 & b_1 \\ a_2 & b_2 \end{pmatrix}.$$

Moreover we define the following two matrices

$$A_x = \begin{pmatrix} c_1 & b_1 \\ c_2 & b_2 \end{pmatrix}$$

and

$$A_y = \begin{pmatrix} a_1 & c_1 \\ a_2 & c_2 \end{pmatrix},$$

observe that the A_x matrix is equal to the A matrix except for the first column (the one referring to the coefficients of x) which has been replaced by the known terms. Similarly the A_y matrix is the same as the A matrix except for the second column (the one referring to the y coefficients) which has been replaced by the known terms. Starting from these three matrices, in particular of their determinants, the system can be solved quickly. In fact, Cramer's method involves calculating the three determinants

$$|A| = \det(A), \qquad |A_x| = \det(A_x),$$
$$|A_y| = \det(A_y),$$

from which

- if $|A| \neq 0$ then the system is determined, i.e., it admits solution, given by

$$x = \frac{|A_x|}{|A|}, \qquad y = \frac{|A_y|}{|A|};$$

- if $|A| = 0$ then the system is impossible or indeterminate, depending on whether $|A_x|$ and $|A_y|$ are also zero or not. In fact, we have

-
  - if also $|A_x| = |A_y| = 0$ then the system is indeterminate;

  - otherwise the system is impossible.

Recall that the determinant of a 2×2 matrix of the type

$$M = \begin{pmatrix} a & b \\ c & d \end{pmatrix}$$

is calculated as

$$\det(M) = ad - bc.$$

Cramer's method is also applied to systems with n equations and n unknowns, using $n \times n$ matrices, whose determinants are increasingly difficult to calculate with increasing n.

3.4.1 Exercise

Solve the following system with Cramer's method

$$\begin{cases} 2x + 4y = 2 - x + 3(y - x) \\ 8x - y = 4 + 3y \,, \end{cases}$$

Solution

We perform the calculations to eliminate the brackets

$$\begin{cases} 2x + 4y = 2 - x + 3y - 3x \\ 8x - y = 4 + 3y \end{cases},$$

we separate the variables from the known terms

$$\begin{cases} 2x + 4y + x - 3y + 3x = 2 \\ 8x - y - 3y = 4 \end{cases}$$

and we write the system in normal form, adding the similar terms,

$$\begin{cases} 6x + y = 2 \\ 8x - 4y = 4 \end{cases}, \quad \begin{cases} 6x + y = 2 \\ 2x - y = 1 \end{cases}$$

The coefficient matrix is as follows

$$A = \begin{pmatrix} 6 & 1 \\ 2 & -1 \end{pmatrix},$$

hence the determinant

$$|A| = 6(-1) - 1(2) = -6 - 2 = -8.$$

The system is determined, it admits solution, being $|A| \neq 0$. We calculate the determinants of the two matrices

$$A_x = \begin{pmatrix} 2 & 1 \\ 1 & -1 \end{pmatrix}, \quad A_y = \begin{pmatrix} 6 & 2 \\ 2 & 1 \end{pmatrix}.$$

We have

$$|A_x| = 2(-1) - 1 = -2 - 1 = -3,$$
$$|A_y| = 6(1) - 2(2) = 6 - 4 = 2,$$

from which the solution of the system

$$x = \frac{|A_x|}{|A|} = \frac{-3}{-8} = \frac{3}{8},$$
$$y = \frac{|A_y|}{|A|} = \frac{2}{-8} = -\frac{1}{4}.$$

The second degree equations

4.1 Generic form

A second degree equation can be reduced to the following general form

$$ax^2 + bx + c = 0\,,$$

with $a, b, c \in \mathbb{R}$ constants and $a \neq 0$.

In general, the coefficients b and c can be zero and we can distinguish the following special cases:

- **pure equation:** if $b = 0$ and $c \neq 0$. The pure equation has the form

$$ax^2 + c = 0\,,$$

if a and c have different signs then the equation admits two distinct real solutions given by

$$x = \pm\sqrt{-\frac{c}{a}}\,,$$

while it does not admit real solutions if a and c have the same sign;

- **spurious equation:** if $b \neq 0$ and $c = 0$. The spurious equation has the form

$$ax^2 + bx = 0\,,$$

and always admits two real solutions, one of which is always zero, given by

$$x_1 = 0, \qquad x_2 = -\frac{b}{a};$$

- **monomial equation:** if $b = 0$ and $c = 0$. The monomial equation has the form

$$ax^2 = 0$$

and admits zero as the only solution.

The general resolution method that we will show shortly has general validity and also includes the solutions of the particular cases set out above.

4.2 Solution method

A second degree equation has at most 2 real solutions (it always has exactly 2 complex ones).

Given the equation

$$ax^2 + bx + c = 0,$$

we define the discriminant of the equation as

$$\Delta = b^2 - 4ac.$$

We have the following cases:

- if $\Delta > 0$ there are two distinct real solutions, given by

$$x_{1,2} = \frac{-b \pm \sqrt{\Delta}}{2a};$$

- if $\Delta = 0$ we have two coincident real solutions, given by

$$x_1 = x_2 = -\frac{b}{2a};$$

- if $\Delta < 0$ the equation does not admit real solutions.

4.2.1 Exercise

Solve the following second degree equation

$$2x(x-2) = 1 - 3x + x^2 \,.$$

Solution

We write the equation in its normal form

$$2x^2 - 2x = 1 - 2x + x^2 \,,$$
$$2x^2 - 2x - x^2 - 1 + 3x = 0 \,, \,,$$

from which

$$x^2 + x - 1 = 0 \,.$$

The discriminant is

$$\Delta = (1)^2 - 4(1)(-1) = 1 + 4 = 5 \,,$$

from which, being $\Delta > 0$, the two distinct real solutions

$$x_{1,2} = \frac{-1 \pm \sqrt{5}}{2}\,.$$

The fractional equations

A fractional equation can be solved following this scheme:

- decompose all possible expressions;

- impose the conditions of existence: all denominator quantities must be different from zero;

- transformation into an equivalent integer equation (under the conditions of existence of the previous point) through *l.c.m.* (least common multiple);

- solve the obtained integer equation, whose solutions can be accepted only if they satisfy the conditions of existence.

5.0.1 Exercise

Solve the following divided equation

$$\frac{2x-3}{x-1} - \frac{2x}{x+1} = \frac{4}{x^2-1}.$$

Solution

We start by decomposing the denominator on the second member

$$\frac{2x-3}{x-1} - \frac{2x}{x+1} = \frac{4}{(x+1)(x-1)},$$

the conditions of existence are obtained by placing

$$x+1 \neq 0, \qquad x \neq -1$$

and

$$x - 1 \neq 0, \qquad x \neq 1.$$

We calculate the *l.c.m.*

$$l.c.m. = (x+1)(x-1),$$

from which

$$\frac{(2x-3)(x+1) - 2x(x-1)}{(x+1)(x-1)} = \frac{4}{(x+1)(x-1)},$$
$$\frac{2x^2 + 2x - 3x - 3 - 2x^2 + 2x}{(x+1)(x-1)} = \frac{4}{(x+1)(x-1)}.$$

By simplifying and performing the calculations, we arrive at the equivalent integer equation (under the conditions of existence)

$$2x - 3x - 3 + 2x = 4,$$
$$x - 3 = 4,$$

from which the solution

$$x = 7,$$

which is acceptable, because it respects the conditions of existence, i.e. $x \neq \pm 1$.

The intervals

6.1 Definition

Intuitively, an interval of $\mathbb{R}$, with endpoints a and b, with $a < b$, is defined as the set made up of all real numbers between a and b.

There are various cases, the interval may in fact include or not include the endpoints.

The intervals are closely related to the inequalities, in fact often the inequalities admit solutions belonging to a certain interval of real numbers.

6.2 Types

Given a set A we have the following definitions

- **Open interval with endpoints** a, b

    An open interval with endpoints a, b, with $a < b$, for a set A is formed by all the numbers between a and b excluding endpoints and is denoted by $]a, b[$, we have:

    $$]a, b[= \{x \in A : a < x < b\}$$

- **Closed interval with endpoints** a, b

    A closed range with endpoints a, b, with $a < b$, for a set A is made up of all the numbers between a and b including endpoints (obviously if $a, b \in A$) and denoted by $[a, b]$, we have:

    $$[a, b] = \{x \in A : a \leq x \leq b\}$$

- **Interval with endpoints** a, b, **left-open**

    A left-open interval with endpoints a, b, with

$a < b$, *for a set A is made up of all numbers between a and b, including b (obviously if $b \in A$) with the exclusion of a and is indicated with $]a, b]$ (or $(a, b]$), we have:*

$$]a, b] = \{x \in A : a < x \leq b\}$$

- **Interval with endpoints a, b, right-open**
  A right-open interval with endpoints a, b, with $a < b$, for a set A is made up of all numbers between a and b, including a (obviously if $a \in A$) with the exclusion of b and is indicated with $[a, b[$ (or $[a, b)$), we have:

  $$[a, b[= \{x \in A : a \leq x < b\}$$

The range representing all real numbers less than $a \in \mathbb{R}$ is written as

$$]-\infty, a[,$$

where the infinity symbol ∞ was used.

Similarly, to indicate the range of all real numbers

greater than a, we write

$$]a, +\infty[.$$

Recall that, concerning the field of reals, the infinity symbol must never be included in the interval and therefore the corresponding square bracket must always be oriented in an appropriate way, such as to exclude it.

Numerical inequalities

7.1 Definition

The following symbols are used for inequalities

$$<, \quad >, \quad \leq, \quad \geq,$$

which express a relation between numbers. These symbols can be read respectively: "less than", "greater than", "less than or equal to" and "greater than or equal to".

For example by writing

$$2 < 5,$$

we mean that the number 2, called the first member (because it is placed to the left of the < symbol), is less than the number 5, called the second member (because it is placed to the right of the < symbol).

It is said that two inequalities have the same direction if the symbol that expresses the relation is the same, i.e., it appears oriented in the same way. For example the two inequalities

$$2 < 5, \qquad 8 < 9,$$

have the same direction, while the two inequalities

$$3 < 4, \qquad 4 > 1,$$

don't have the same direction.

7.2 Property

Inequalities have some properties, we list the five basic properties below.

- **First property.** *Given an inequality, adding or subtracting any number to both sides does not change its direction. We can write:*

$$a \underbrace{<}_{>,\,\leq,\,\geq} b \implies a+c \underbrace{<}_{>,\,\leq,\,\geq} b+c, \ \forall\, c \in \mathbb{R};$$

- **Second property.** *Given an inequality, multiplying or dividing both sides by any non-zero number, its direction does not change if this number is positive, it changes if it is negative. We can write:*

$$a \underbrace{<}_{>,\,\leq,\,\geq} b \implies$$

$$\begin{cases} a \cdot c \underbrace{<}_{>,\,\leq,\,\geq} b \cdot c, & \forall c \in \mathbb{R},\ c > 0 \\ a \cdot c \underbrace{>}_{<,\,\geq,\,\leq} b \cdot c, & \forall c \in \mathbb{R},\ c < 0 \end{cases} ;$$

- **Third property.** *Given an inequality composed of non-zero numbers, if the numbers are discordant the inequality of the reciprocals has the same direction, otherwise the opposite direction. We can write:*

$$a \underbrace{<}_{>,\,\leq,\,\geq} b \implies \begin{cases} \frac{1}{a} \underbrace{<}_{>,\,\leq,\,\geq} \frac{1}{b}, & a, b \text{ discordant} \\ \frac{1}{a} \underbrace{>}_{<,\,\geq,\,\leq} \frac{1}{b}, & a, b \text{ concordant} \end{cases}$$

- **Fourth property.** *Given two inequalities of the same direction, adding the homolo-*

gous members we obtain an inequality of the same verse. We can write:

$$\begin{cases} a \underset{>,\,\leq,\,\geq}{<} b \\ c \underset{>,\,\leq,\,\geq}{<} d \end{cases} \implies a+c \underset{>,\,\leq,\,\geq}{<} b+d\,;$$

- **Fifth property.** *Given two inequalities of the same direction and with positive members, multiplying the homologous members we obtain an inequality of the same verse. In symbols we can write:*

$$\begin{cases} a \underset{>,\,\leq,\,\geq}{<} b,\ a,b>0 \\ c \underset{>,\,\leq,\,\geq}{<} d,\ c,d>0 \end{cases} \implies a\cdot c \underset{>,\,\leq,\,\geq}{<} b\cdot d\,.$$

Inequalities

8.1 Definition

In a generic inequality one or more unknowns appear. The simplest inequalities have only one unknown, usually denoted by the letter x.

Solving an inequality means finding the set of variability of the unknown x, or understanding which are the values of x that satisfy the inequality.

8.2 Types

Similarly to equations, inequalities can also be divided into various groups, depending on the case. We have:

- **1st degree inequality**

    An inequality is said to be of the first degree or linear if the unknown appears only at the first degree (x);

- **2nd degree inequality**

    An inequality is said to be of the second degree if the maximum power to which the unknown is raised is two (x^2);

- **inequality of degree $n-$ th**

    An inequality is said of degree n-th if the maximum power to which the unknown is raised is n (x^n);

- **numerical inequality**

An inequality is called numerical if no other letters appear beyond the unknown;

- **literal inequality**

   An inequality is said to be literal if other letters appear besides the unknown;

- **integer inequality**

   An inequality is said to be integer if the unknown appears only in the numerator of any fractions;

- **fractional inequality**

   An inequality is said to be fractional if there are fractions and the unknown also appears in the denominator.

8.3 Principles of equivalence

Two inequalities are called **equivalent** if they have the same solution set.

As for the equations, also for the inequalities the

simplification steps must respect the rules contained in the following equivalence principles:

- **First principle of equivalence**

  Given an inequality, adding or subtracting any expression (or number) to both sides gives an equivalent equation. In symbols we write:

  $$f(x) \underbrace{<}_{>,\,\leq,\,\geq} a \implies$$
  $$f(x) + c(x) \underbrace{<}_{>,\,\leq,\,\geq} a + c(x)\,, \quad \forall\, c(x)\,;$$

- **Second principle of equivalence**

  Given an inequality, multiplying or dividing both sides by any positive expression (or number) yields an equivalent inequality. On the other hand multiplying or dividing both sides by any negative expression (or number) and changing the direction, we ob-

tain an equivalent inequality. In symbols we write:

$$f(x) \underset{>,\ \leq,\ \geq}{<} a \implies$$

$$\begin{cases} f(x) \cdot c(x) \underset{>,\ \leq,\ \geq}{<} a \cdot c(x)\,,\ \forall\, c(x) > 0 \\ f(x) \cdot c(x) \underset{<,\ \geq,\ \leq}{>} a \cdot c(x)\,,\ \forall\, c(x) < 0 \end{cases}.$$

Note that the first principle is what justifies "carrying" a term from one member to another by changing its sign.

The second principle, on the other hand, justifies the change of the sign of all the terms of an inequality, remembering, however, that its direction must also be changed at the same time. In fact, this operation is equivalent to multiplying both members by the negative number (-1).

The first degree inequalities

First degree inequalities are also called linear inequalities.

9.1 Integer inequalities

An example of first degree integer inequality is the following

$$3(x-2) + \frac{1}{4} > -x + \frac{x}{3}. \qquad (9.1)$$

To solve it we use the second principle of equivalence and multiply both members by the *l.c.m.*

of the denominators, i.e.

$$l.c.m.(3, 4) = 12 > 0,$$

without changing the direction of the inequality being positive, so as to eliminate the fractions

$$\left(3(x-2) + \frac{1}{4}\right) \cdot 12 > \left(-x + \frac{x}{3}\right) \cdot 12,$$

from which

$$3 \cdot 12(x-2) + \frac{12}{4} > -12x + \frac{12x}{3}$$

and

$$36(x-2) + 3 > -12x + 4x.$$

We eliminate the brackets and add the similar terms, obtaining

$$36x - 72 + 3 > -8x,$$

from which

$$36x - 69 > -8x.$$

We use the first principle of equivalence, adding the expression

$$8x + 69$$

to both members, getting

$$36x - 69 + 8x + 69 > -8x + 8x + 69,$$

from which, adding similar terms,

$$44x > 69.$$

Finally, we use the second principle again by dividing both members by the number 44 to get the following inequality, equivalent to the one given in Eq. (9.1),

$$x > \frac{69}{44},$$

which immediately provides the solution set, the one to which the unknown x must belong to satisfy the inequality of Eq. (9.1). In fact the solution set is

$$S = \left\{ x \in \mathbb{R} : x > \frac{69}{44} \right\},$$

which can be written, using interval notation, such as

$$S = \left] -\infty, \frac{69}{44} \right[.$$

In general, to solve a linear integer inequality one must perform the simplification steps necessary to obtain other equivalent inequalities, using the two principles of equivalence, trying to isolate the unknown in the first member, until the solution set is obtained.

9.2 Integer literal inequalities

The inequalities that contain other letters besides the unknown are said literal inequalities. These letters are sometimes called parameters and must be considered to all intents and purposes as if they were fixed numbers, without however knowing their value.

It is necessary to perform a discussion on the parameters to solve the inequality and to correctly use the equivalence principles or the general rules. For example, consider the literal integer linear inequality

$$a \cdot x > b,$$

where x is the unknown and a and b are real parameters. In this case it might be tempting to isolate the unknown x to the first member, dividing both members by a, however this operation requires knowledge of the sign of the parameter

a to decide whether to change the direction of the inequality, or discuss the special case where a was zero. In general, there are three cases: a can be greater than, equal to or less than zero. It is necessary to discuss the literal inequality, which leads to the following scheme:

$$\underbrace{a = 0}_{} \qquad \underbrace{a > 0}_{x > \frac{b}{a}} \quad \underbrace{a < 0}_{x < \frac{b}{a}}$$

$$\underbrace{0 \cdot x > b}_{}$$

$$\underbrace{b < 0}_{\forall x \in \mathbb{R}} \quad \underbrace{b \geq 0}_{\nexists x \in \mathbb{R}}$$

9.2.1 Exercise

Solve the following literal integer inequality with the parameter k

$$(k-2)(x+3) + 5x < 2 + 3kx - k\,.$$

Solution

We can calculate

$$kx - 2x + 3k - 6 + 5x < 2 + 3kx - k,$$

we take all the terms containing the unknown x to the first member and the others to the second member, using the first equivalence principle once or more,

$$kx - 2x + 5x - 3kx < 2 - k - 3k + 6,$$

we add similar terms and collect at first member the quantities with x,

$$(3 - 2k)x < 8 - 4k.$$

Now begins the discussion on the parameter, in fact we cannot divide by $(3-2k)$ if this were zero and we must change the direction if it were negative as stated by the second principle of equivalence. So we have the following scheme

$$\underbrace{\underbrace{3-2k=0}_{k=\frac{3}{2}}}_{\underbrace{0\cdot x < 8-4\left(\frac{3}{2}\right)}_{\underbrace{0\cdot x < 2}_{\forall x \in \mathbb{R}}}} \qquad \underbrace{3-2k>0}_{\underbrace{k<\frac{3}{2}}_{x<\frac{8-4k}{3-2k}}} \qquad \underbrace{3-2k<0}_{\underbrace{k>\frac{3}{2}}_{x>\frac{8-4k}{3-2k}}}$$

9.2.2 Exercise

Solve the following literal integer inequality with the parameter k

$$3(k-x)+5x \geq k(x+2).$$

Solution

After some algebraic calculations we obtain

$$3k-3x+5x \geq kx+2k.$$

We add the similar terms and bring to the first member all quantities containing the unknown and to the second member all the rest, using the first principle of equivalence:

$$3k + 2x > kx + 2k\,,$$

$$2x - kx \geq 2k - 3k\,,$$

$$x(2-k) \geq -k\,.$$

Finally, the discussion on the parameter leads to the following solution scheme

$\underbrace{2-k=0}$	$\underbrace{2-k>0}$	$\underbrace{2-k<0}$
$\underbrace{k=2}$	$\underbrace{k<2}$	$\underbrace{k>2}$
$\underbrace{0\cdot x < -2}$	$x \geq \frac{-k}{2-k}$	$x \leq \frac{-k}{2-k}$
$\nexists x \in \mathbb{R}$		

9.3 Systems of inequalities

A system of inequalities is made up of several inequalities and solving it means finding the common solution to all the inequalities. It is therefore necessary to solve the various inequalities and then performing the so-called intersection of the various solution sets found. This means that the values of x that satisfy the system are only those that simultaneously satisfy all the inequalities that compose it. Consider, as an example, the system of two inequalities given by

$$\begin{cases} 2x + 3 < 0 \\ x - 2 > 0 \end{cases}.$$

We start by solving the first inequality

$$2x + 3 < 0,$$
$$2x < -3,$$
$$x < -\frac{3}{2}.$$

The solution set is

$$S_1 = \left]-\infty, -\frac{3}{2}\right[.$$

The second inequality of the system leads to

$$x - 2 > 0,$$
$$x > 2,$$

hence the solution set

$$S_2 =]2, \infty[.$$

We can represent these two sets in the real line in this way

THE FIRST DEGREE INEQUALITIES

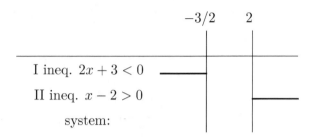

In the diagram, the presence of a black line means that the x in that interval satisfy the respective inequality. Given a certain interval, the system will admit a solution in that interval only if the lines of all the inequalities are present. In this case in the diagram above, for example, between $-\infty$ and $-3/2$ there is only the line of the first inequality, but not that of the second and therefore the system will have no solution for the x in that interval. Finally, the system has the empty set as solution (it admits no solutions) and we write

$$S = S_1 \cap S_2 = \{\}.$$

9.3.1 Exercise

Solve the following system of inequalities

$$\begin{cases} 6x + 3 \leq 2(x + 5) \\ -x + 1 > 2x \\ -2x + 3 \geq 8(1 - x) \\ x < -2 - x \end{cases}$$

Solution

We must first solve the four inequalities and find the four solution sets. For the first one we have

$$6x + 3 \leq 2x + 10,$$
$$6x - 2x \leq 10 - 3,$$
$$4x \leq 7,$$

from which

$$x \leq \frac{7}{4}$$

and its whole solution is

$$S_1 = \left]-\infty, \frac{7}{4}\right] = \left]-\infty, 1.75\right].$$

The second inequality leads to

$$-x + 1 > 2x,$$
$$-x - 2x > -1,$$
$$-3x > -1,$$
$$3x < 1,$$

from which

$$x < \frac{1}{3},$$

and hence

$$S_2 = \left]-\infty, \frac{1}{3}\right[.$$

For the third inequality we have

$$-2x + 3 \geq 8 - 8x\,,$$
$$-2x + 8x \geq 8 - 3\,,$$
$$6x \geq 5\,,$$

from which

$$x \geq \frac{5}{6}\,,$$

$$S_3 = \left[\frac{5}{6}, \infty\right[\,.$$

The last inequality, on the other hand, gives

$$x < -2 - x\,,$$
$$x + x < -2\,,$$
$$2x < -2\,,$$

from which

$$x < -1\,,$$

hence the solution set

$$S_4 = \,]-\infty, -1[\,.$$

To find the intersection between the four solution sets, i.e.

$$S = S_1 \cap S_2 \cap S_3 \cap S_4,$$

representing the solution set of the system, we use the following scheme:

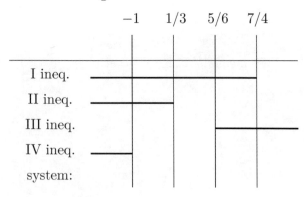

where we used the approximations:

$$\frac{1}{3} \simeq 0.33, \qquad \frac{5}{6} \simeq 0.83, \qquad \frac{7}{4} = 1.75.$$

Again, for the system we have

$$S = S_1 \cap S_2 \cap S_3 \cap S_4 = \{\},$$

in fact there is no interval or point in which all four inequalities of the system have a common solution.

Sign of products or ratios

10.1 Sign of a product

Studying the sign of a product of two or more polynomials means calculate when the product is positive and when it is negative.

Given two real numbers, their product will be positive if they have the same sign, negative if they have different signs. This continues to be true even when monomials or polynomials are multiplied. In particular, consider the product

$$(2x-3)(x+5),$$

this will therefore be positive if both factors are positive or negative (same sign) and negative if they have different signs. It is therefore advisable to study the sign of the two factors separately, for example by looking for when they are positive. In fact, it will be immediate to understand when they are null or negative. Therefore we write

$$2x-3 > 0,$$

which has the solution

$$x > \frac{3}{2}.$$

The first factor is positive for $x > 3/2$, zero for $x = 3/2$ and negative if $x < 3/2$.

SIGN OF PRODUCTS OR RATIOS

For the second factor we write

$$x + 5 > 0,$$

from which

$$x > -5.$$

The second factor is therefore positive for $x > -5$, zero for $x = -5$ and negative if $x < -5$.

We can summarize the behavior of the two factors in the following table:

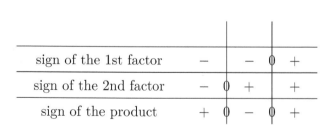

In this table we have placed the symbols $(-, +, 0)$ depending on whether the sign of the factor of that row is positive, negative or null. The product sign is given by the product of these symbols in the boxes above, in the usual way, i.e.,

$$(+) \cdot (+) = (+), \quad (-) \cdot (+) = (-),$$
$$(-) \cdot (+) = (-), \quad (-) \cdot (-) = (+).$$

In this case we see that the product

$$(2x - 3)(x + 5)$$

is positive for

$$x < -5,$$

or

$$x > 3/2,$$

is negative for

$$-5 < x < 3/2$$

and is zero when

$$x = -5,$$

or

$$x = 3/2.$$

If we were to solve the inequality

$$(2x - 3)(x + 5) > 0,$$

the study of the sign of the product just made would allow us to find the solution set which is written as a union of two intervals in this way

$$S =]-\infty, -5[\cup \left]\frac{3}{2}, +\infty\right[,$$

or, equivalently,

$$S = \left\{ x \in \mathbb{R} : x < -5 \vee x > \frac{3}{2} \right\}.$$

The $\vee$ symbol used is a logical symbol that means

"or", while the logical symbol ∧ means "and".

10.1.1 Exercise

Solve the following inequality

$$(x-2)(3x+1)(5-2x) \leq 0.$$

Solution

This inequality can be solved by studying the sign of the three factors, using the strategies seen so far.

For the sign of the first factor we solve the inequality

$$x - 2 > 0,$$

which leads to

$$x > 2.$$

The first factor, $(x-2)$, is positive for $x > 2$, zero for $x = 2$, and negative for $x < 2$. For the second factor we write

$$3x + 1 > 0,$$

or

$$3x > -1,$$

from which

$$x > \frac{1}{3}.$$

The second factor $(3x + 1)$ is therefore positive for $x > -1/3$, zero for $x = -1/3$ and negative for $x < -1/3$. Finally for the third factor

$$5 - 2x > 0,$$

from which

$$-2x > -5$$

SIGN OF PRODUCTS OR RATIOS

and

$$x < \frac{5}{2}.$$

The third factor, $(5 - 2x)$, is positive for $x < 5/2$, zero for $x = 5/2$ and negative for $x > 5/2$.

We can summarize the behavior of the three factors in the following table:

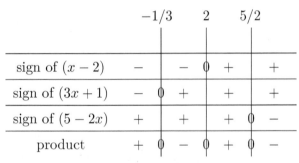

From the table it is clear that the solution set of the initial inequality is

$$S = \left[-\frac{1}{3}, 2\right] \cup \left[\frac{5}{2}, +\infty\right[.$$

10.2 Fractional inequalities

The study of the sign of a product is also used for the resolution of fractional inequalities, i.e., an inequality where the unknown appears as a denominator at least once. In fact, the sign of the ratio between two expressions follows the same rules as the sign of their product, with the only care that the denominator cannot be zero (condition of existence). The table used previously for the study of the sign of a product will be similar in the case of the ratio, it will only be necessary to replace the various 0 written where the denominator is zero with the $\nexists$ symbol, which means "does not exist ". This is done to remember that for that value, or those values, the denominator is not defined and therefore neither is the ratio. To give an example, suppose we want to solve the following fractional inequality

$$\frac{8x+2}{2-15x} > 0.$$

A study of the sign of the numerator and denominator taken separately is performed and a study of the sign of the ratio is done with a table similar to that for the product.

We study the sign of the numerator

$$8x + 2 > 0,$$

from which

$$8x > -2,$$
$$x > -\frac{1}{4}.$$

We therefore know that the numerator $(8x + 2)$ is positive for $x > -1/4$, zero for $x = -1/4$ and negative for $x < -1/4$. For the denominator we

SIGN OF PRODUCTS OR RATIOS

have

$$2 - 15x > 0,$$

or

$$-15x > -2,$$

from which

$$x < \frac{2}{15}.$$

The denominator $(2 - 15x)$ is positive for $x < 2/15$, zero for $x = 2/15$ and negative for $x > 2/15$.

The scheme in this case is as follows:

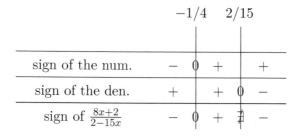

We see that where the denominator is zero, the ratio does not exist, it has no meaning, and therefore, in the diagram, the $\nexists$ symbol has been placed under the appropriate value, i.e. $2/15$. This is the only difference with the study of the sign of the product between numerator and denominator. Finally, the solution is

$$S = \left]-\infty, -\frac{1}{4}\right[\cup \left]\frac{2}{15}, +\infty\right[.$$

To conclude, in order to solve a fractional inequality we must first obtain a relation between a quantity and zero, of the type

$$\frac{N(x)}{D(x)} \underset{<,\ \geq,\ \leq}{>} 0,$$

where $N(x)$ and $D(x)$ are the numerator and denominator, functions of x. Subsequently, the study of the sign is performed, as shown previously.

10.3 Fractional literal inequalities

The fractional literal inequalities are fractional inequalities where other letters appear in addition to the unknown and must be treated with an appropriate discussion as in the case of literal integer inequalities.

The second degree inequalities

11.1 Generic form

A second degree inequality has, in general, the following form

$$ax^2 + bx + c \underbrace{>}_{<,\ \geq,\ \leq} 0\,,$$

with $a, b, c \in \mathbb{R}$ constants and $a \neq 0$.

11.2 Solution method

In general, to find the solutions, we first solve its associated equation, i.e.

$$ax^2 + bx + c = 0,$$

which has the discriminant

$$\Delta = b^2 - 4ac.$$

If $\Delta > 0$ the two distinct solutions of the equation are

$$x_1 = \frac{-b + \sqrt{\Delta}}{2a}$$

and

$$x_2 = \frac{-b - \sqrt{\Delta}}{2a}.$$

If $\Delta = 0$ we have two coincident solutions

$$x_1 = x_2 = -\frac{b}{2a},$$

whereas if $\Delta < 0$ there are no real solutions.

Let us consider, on a system of Cartesian axes (x, y), the plot of the function

$$y = ax^2 + bx + c,$$

if, as hypothesized $a \neq 0$, it will be given by a parabola with concavity pointing upwards or downwards depending on the sign of a. The intersections with the abscissa axis is given by the solutions of the second degree equation

$$ax^2 + bx + c = 0.$$

Solve the starting inequality

$$ax^2 + bx + c \underbrace{>}_{<,\ \geq,\ \leq} 0,$$

means calculating for what x happens

$$y \underbrace{>}_{<,\ \geq,\ \leq} 0,$$

it is therefore necessary to determine when the plot of the parabola is above ($>$), below ($<$) the abscissa axis or is above and intersects it ($\geq$) or below and intersects it ($\leq$). In fact, the abscissa axis has the Cartesian equation

$$y = 0.$$

For example, suppose we want to solve the second degree inequality

$$2x^2 - 3x + 1 > 0.$$

The associated equation is

$$2x^2 - 3x + 1 = 0,$$

from which

$$\Delta = 9 - 8 = 1 > 0,$$

the solutions of the equation are

$$x_1 = 1, \qquad x_2 = \frac{1}{2}.$$

Since $a = 2 > 0$, the concavity of the parabola faces upwards and we are interested in the set of x for which the plot is strictly above the abscissa axis (in fact in the inequality there is the direction $>$). Since $x_1 = 1$ and $x_2 = 1/2$ represent the intersections with the abscissa axis then the parabola will be above the abscissa axis if $x < 1/2$ or if $x > 1$. For a better understanding, refer to the plot shown in Figure 11.1.

Therefore the solution set is

$$S = \left]-\infty, \frac{1}{2}\right[\cup \left]1, +\infty\right[.$$

In the case where the associated equation has $\Delta < 0$, i.e. there are no intersections of the parabola with the abscissa axis, depending on the concavity the parabola will always be above the

Figure 11.1: Plot of $y = 2x^2 - 3x + 1$.

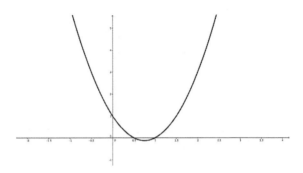

abscissa axis or always below.

11.2.1 Exercise

Solve the second degree inequality

$$-x^2 + x - 1 \leq 0 \,.$$

Figure 11.2: Parabola $y = -x^2 + x - 1$.

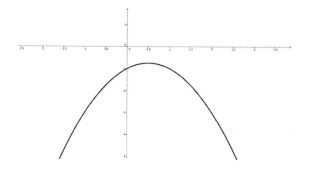

Solution

Its associated equation,

$$-x^2 + x - 1 = 0\,,$$

has negative discriminant, $\Delta < 0$, in fact

$$\Delta = 1 - 4 = -3 < 0$$

and the parabola

$$y = -x^2 + x - 1$$

never intersects the abscissa axis. Its concavity points downwards, being $a = -1 < 0$ and therefore its plot will always be under the abscissa axis, as shown in Figure 11.2. Therefore the inequality

$$-x^2 + x - 1 \leq 0$$

is satisfied for every x real and the solution set is

$$S = \mathbb{R}.$$

The irrational equations

An irrational equation is an equation where the unknown appears under the root sign.

12.1 First case

Let us consider an irrational equation of the type

$$\sqrt{f(x)} = g(x),$$

where $f(x)$ and $g(x)$ are generic functions of x. The solution of this equation is obtained by solving the system

$$\begin{cases} f(x) \geq 0 \\ g(x) \geq 0 \\ f(x) = g^2(x) \end{cases}$$

where the first condition represents the condition of existence for the square root (non-negative argument), the second is due to the fact that $g(x)$, being equal to a square root, is always non-negative and, finally, the last equation represents the starting one with both sides squared (the root is simplified).

We observe that not all these three equations or inequalities are independent, in fact the third condition automatically guarantees the first condition (in fact $f(x) = g^2(x) \geq 0$). Therefore the final system that has to be solved to obtain solution of the irrational equation

$$\sqrt{f(x)} = g(x)$$

is the following

$$\begin{cases} g(x) \geq 0 \\ f(x) = g^2(x) \end{cases}$$

12.2 Second case

Consider an irrational equation of the type

$$\sqrt{f(x)} = \sqrt{g(x)},$$

with similar considerations to the previous case, we have the following solution system

$$\begin{cases} f(x) \geq 0 \\ g(x) \geq 0 \\ f(x) = g(x) \end{cases}.$$

12.2.1 Exercise

Solve the irrational equation

$$\sqrt{x-2} = 5.$$

Solution

In this case, since there is simply a positive number on the second member, we can directly solve the equivalent equation

$$x - 2 = 5^2,$$

from which

$$x = 27.$$

The irrational inequalities

An irrational inequality is an inequality in which the unknown appears under the root sign.

13.1 First case

Consider firstly an irrational inequality of the form

$$\sqrt{f(x)} \underset{<,\,\geq,\,\leq}{>} g(x),$$

where $f(x)$ and $g(x)$ are generic functions of x. The solutions depend on the direction, so we treat the four cases separately.

Consider the inequality

$$\sqrt{f(x)} > g(x),$$

to obtain the solution set we must separate the cases where $g(x) \geq 0$ or $g(x) < 0$, from which

$$\begin{cases} g(x) \geq 0 \\ f(x) > g^2(x) \end{cases} \quad \text{or} \quad \begin{cases} g(x) < 0 \\ f(x) \geq 0 \end{cases},$$

where the second condition in the second system represents the condition of existence for the square root (non-negative argument). Note that the second system means that under the hypothesis $g(x) < 0$ the starting inequality is always satisfied, under the existence condition $f(x) \geq 0$ (a negative number is always less than a square root, if the latter exists).

Once these two systems have been solved, it is necessary to perform the **union** of the respective solution sets to obtain the total solution set of the original inequality.

With similar considerations, for the inequality

$$\sqrt{f(x)} \geq g(x),$$

we have

$$\begin{cases} g(x) \geq 0 \\ f(x) \geq g^2(x) \end{cases} \text{ or } \begin{cases} g(x) < 0 \\ f(x) \geq 0 \end{cases}.$$

Similarly for the inequality

$$\sqrt{f(x)} < g(x),$$

we have

$$\begin{cases} g(x) \geq 0 \\ f(x) \geq 0 \\ f(x) < g^2(x) \end{cases},$$

while for the inequality

$$\sqrt{f(x)} \leq g(x),$$

the system is

$$\begin{cases} g(x) \geq 0 \\ f(x) \geq 0 \\ f(x) \leq g^2(x) \end{cases}$$

Observe that in the last two cases, if $g(x) < 0$ the inequality to be solved does not admit solutions and therefore we have only one system.

13.2 Second case

Let us consider now an irrational inequality of the type

$$\sqrt{f(x)} \underbrace{>}_{<,\, \geq,\, \leq} \sqrt{g(x)},$$

where $f(x)$ and $g(x)$ are generic functions of x. In this case, in order to obtain the solution set, we have to solve the system

$$\begin{cases} f(x) \geq 0 \\ g(x) \geq 0 \\ f(x) \underbrace{>}_{<,\, \geq,\, \leq} g(x) \end{cases},$$

where the first two inequalities guarantee the existence of the two square roots.

13.2.1 Exercise

Solve the irrational inequality

$$\sqrt{x+9} > 3 - x.$$

Solution

We have to solve

$$\begin{cases} 3 - x \geq 0 \\ x + 9 > (3-x)^2 \end{cases} \quad \text{or} \quad \begin{cases} 3 - x < 0 \\ x + 9 \geq 0 \end{cases}.$$

For the first system, we can write

$$\begin{cases} -x \geq -3 \\ x+9 > 9 + x^2 - 6x \end{cases}, \quad \begin{cases} x \leq 3 \\ x > x^2 - 6x \end{cases},$$

from which

$$\begin{cases} x \leq 3 \\ x - x^2 + 6x > 0 \end{cases}, \quad \begin{cases} x \leq 3 \\ x^2 - 7x < 0 \end{cases}.$$

Consider the associated equation

$$x^2 - 7x = x(x-7) = 0,$$

whose solutions are[1]

$$x_1 = 0, \quad x_2 = 7.$$

The parabola

$$y = x^2 - 7x$$

---

[1] a product of several factors is zero if and only if at least one factor is zero.

has concavity turned upwards, so that the solution set of the inequality

$$x^2 - 7x < 0$$

is represented by the interval

$$0 < x < 7\,,,$$

which can also be written as

$$]0, 7[\,.$$

We intersect the solution sets of the two inequalities of the first system with the help of the following scheme:

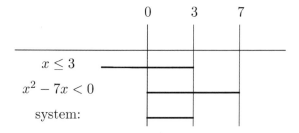

so the solution of the first system is

$$S_1 =]0, 3].$$

Consider now the second system

$$\begin{cases} 3 - x < 0 \\ x + 9 \geq 0 \end{cases},$$

we have

$$\begin{cases} -x < -3 \\ x \geq -9 \end{cases}, \qquad \begin{cases} x > 3 \\ x \geq -9 \end{cases}.$$

The scheme for the intersection is as follows

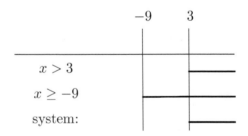

THE IRRATIONAL INEQUALITIES

hence the solution set of the second system

$$S_2 =]3, +\infty[.$$

We must calculate the union of the two solution sets S_1 and S_2. We use the following scheme:

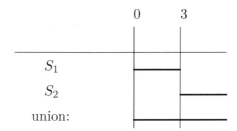

where in this case of union, instead of intersection, the values of the first or second set are taken, i.e. the last line appears if there is at least one line in that interval, from any equation. Finally, the solution of the inequality

$$\sqrt{x+9} > 3 - x$$

is

$$S = S_1 \cup S_2 =]0, +\infty[.$$

13.2.2 Exercise

Solve the irrational inequality

$$3x - 7 + \sqrt{9x^2 - 16} < 0.$$

Solution

We write the inequality in the form

$$\sqrt{9x^2 - 16} < -3x + 7,$$

we must therefore solve the system

$$\begin{cases} -3x + 7 \geq 0 \\ 9x^2 - 16 \geq 0 \\ 9x^2 - 16 < (-3x + 7)^2 \end{cases}$$

The first inequality gives

$$-3x \geq -7,$$
$$x \leq \frac{7}{3},$$

hence the solution set

$$S_1 = \left]-\infty, \frac{7}{3}\right].$$

For the second inequality of the system we consider the associated equation

$$9x^2 - 16 = 0,$$

whose solutions are

$$9x^2 = 16,$$
$$x_{1,2} = \pm \frac{4}{3}.$$

The parabola

$$y = 9x^2 - 16$$

has the concavity facing upwards, so the solutions of the second inequality of the system are

$$x \leq -\frac{4}{3} \quad \text{o} \quad x \geq \frac{4}{3},$$

i.e.

$$S_2 = \left[-\infty, -\frac{4}{3}\right] \cup \left[\frac{4}{3}, +\infty\right].$$

For the third inequality of the system,

$$9x^2 - 16 < (-3x + 7)^2,$$

we calculate

$$9x^2 - 16 < 9x^2 + 49 - 42x,$$
$$-16 + 42x < 49$$
$$42x < 65$$

from which

$$x < \frac{65}{42}$$

and

$$S_3 = \left]-\infty, \frac{65}{42}\right[\, .$$

To completely solve the starting inequality, i.e.

$$3x - 7 + \sqrt{9x^2 - 16} < 0 \, ,$$

we intersect the three solution sets through the scheme:

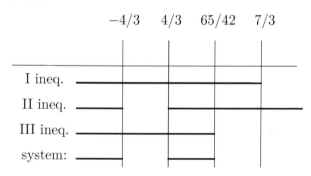

where we used the approximations

$$\frac{4}{3} \simeq 1.33 \, , \qquad \frac{65}{42} \simeq 1.55 \, , \qquad \frac{7}{3} \simeq 2.33 \, .$$

therefore the solution set of the inequality is

$$S = S_1 \cup S_2 \cup S_3 = \left[-\infty, -\frac{4}{3}\right] \cup \left[\frac{4}{3}, \frac{65}{42}\right[\, .$$